LIFE
STORY

ERIC MADDERN

illustrations by

LEO DUFF

BARRON'S

Long ago when the world was new,
the land was bare and the only living things
were tiny, tiny specks, floating about in the sunlit sea.
Millions and millions and millions of years went by
and the little specks kept growing and splitting,
slowly changing into all sorts of new kinds,
until at last some came together
and made the first living cells, still too small to see.

Some cells made their food using light from the sun;
they were the first green plants.
Other cells got their food by eating up the green ones;
they were the first animals.

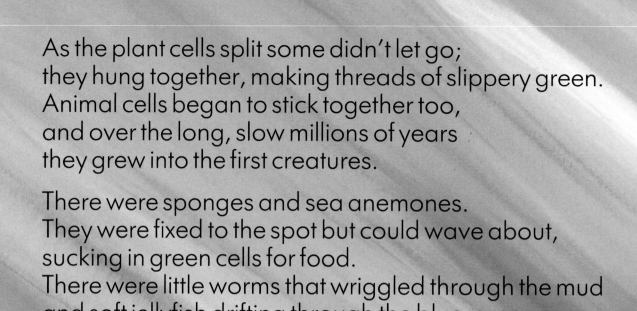

As the plant cells split some didn't let go;
they hung together, making threads of slippery green.
Animal cells began to stick together too,
and over the long, slow millions of years
they grew into the first creatures.

There were sponges and sea anemones.
They were fixed to the spot but could wave about,
sucking in green cells for food.
There were little worms that wriggled through the mud
and soft jellyfish drifting through the blue-green sea.

Millions of years passed and, bit by bit,
life began to get hard under the salty sea.
The big creatures started to eat the smaller ones.
There was sharp-shelled nautilus,
the first with eyes to see, that jerked around
catching shrimps and worms with its tentacles;
and the armor-plated sea scorpion
that crawled about, snapping its deadly claws.
But even these fierce creatures couldn't eat
snails, safe inside their hard shells,
starfish with tough, spiny skins
or anemones hiding in the coral.

Some of the worm-like creatures
learned to get away by growing hard parts inside.
With bendy backbones and bony heads
they wiggled along above the seabed,
sucking up little bits of food.

A few went all the way to rivers and lakes
and there, as time went by,
they slowly grew big jaws and clever fins.
When they came back to the sea
they could open their mouths so wide
that they swallowed all the little ones up.
With their swishing tails they swam fast and free,
the first of the silvery fish.

By now the sea was full of life
but the land was still quite bare,
until slippery green plants began to grow out of the water
and slowly covered the rocks with carpets of moss.
After millions of years there were forests of ferns
and the land was leafy and green.

Then out of the sea came small, scaly worms
crawling on lots of little legs, to eat and live on the leaves.
In a few more million years
there were beetles, spiders and flies,

and flitting about in the air,
the beautiful dragonfly.
In the warm swamps even a few fish
started coming up to gulp for air.

For these fish in the swamps everything was fine
until the weather started to change.
For thousands of years it got hotter and hotter,
the lakes and swamps dried out,
and many fish died in the mud.
But the ones that could already gulp down air
dragged themselves across the ground
to the nearest pool
and lived to lay eggs as before.
Their babies still hatched out in the water,
but they learned to move about on the land.
They were the first amphibians.

A few amphibians began exploring the land,
moving out into the drier places.
But here they were in danger from the heat of the sun
so some grew tough skins to keep them from drying out
and began to lay eggs with hard shells
keeping little pools of water safe inside.

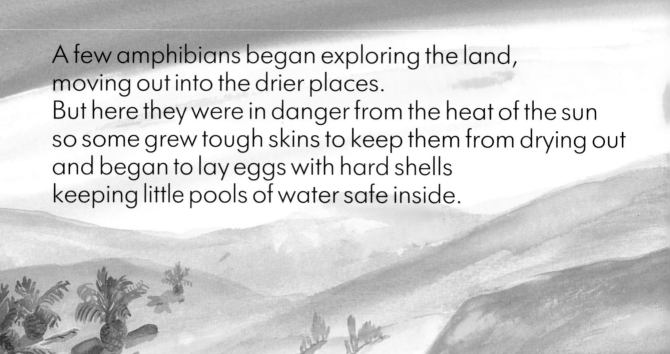

When the tiny babies hatched
they were all ready to bite and run;
they didn't need to start life in the water.
These creatures were the first reptiles;
they could live anywhere on the land.

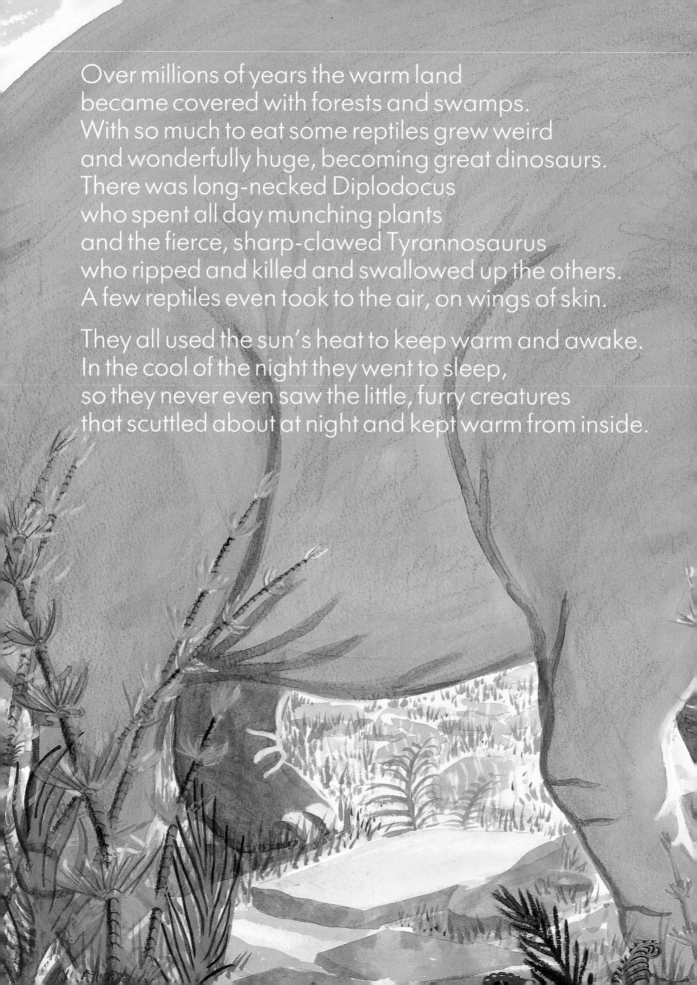

Over millions of years the warm land
became covered with forests and swamps.
With so much to eat some reptiles grew weird
and wonderfully huge, becoming great dinosaurs.
There was long-necked Diplodocus
who spent all day munching plants
and the fierce, sharp-clawed Tyrannosaurus
who ripped and killed and swallowed up the others.
A few reptiles even took to the air, on wings of skin.

They all used the sun's heat to keep warm and awake.
In the cool of the night they went to sleep,
so they never even saw the little, furry creatures
that scuttled about at night and kept warm from inside.

Up in the trees some light-boned reptiles
were finding new ways to move in the air.
Over millions of years the ones who did well
grew long wing feathers to help them fly high,
fluffy feathers for warmth and strong pecking beaks.
They were the first of the flying birds.

It was around this time, too, that the first flowers grew
and from them came fruits and seeds.
The bees buzzed around sipping nectar from the flowers,
and the birds ate berries and sang.

All was well in the warm green world.
Then suddenly disaster struck. . .

Some say a big rock from outer space
came crashing into the Earth,
making volcanoes spout and fume
and blocking out the sun's warmth with dust.
It grew cold and the dinosaurs were in terrible trouble.
Without the sun they couldn't wake up,
and when it stayed cold for months and months
all the dinosaurs died.
But the small creatures that kept warm from inside
didn't mind the cold.
They were the first mammals,
and suddenly the world was all theirs.

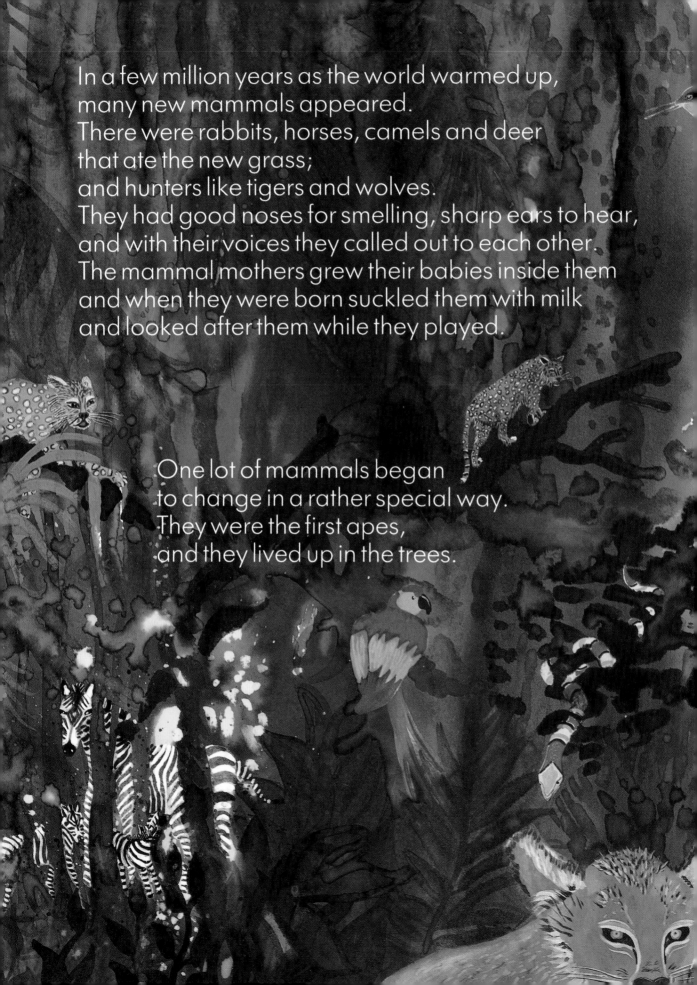

In a few million years as the world warmed up,
many new mammals appeared.
There were rabbits, horses, camels and deer
that ate the new grass;
and hunters like tigers and wolves.
They had good noses for smelling, sharp ears to hear,
and with their voices they called out to each other.
The mammal mothers grew their babies inside them
and when they were born suckled them with milk
and looked after them while they played.

One lot of mammals began
to change in a rather special way.
They were the first apes,
and they lived up in the trees.

As they swung through the forests
looking for fruit
the apes chattered together.

Their hands grew strong
from always holding on,
their eyes grew sharp from looking out,
and the little ones rode on their mother's backs,
learning all the time.
A few bigger ape children played at the forest edge
and sometimes practiced walking on their two hind legs.
This way they could stand tall, see food from far away
and watch out for danger.
Now their hands were free to carry things,
throw stones and poke with sticks.
They slowly discovered a new life on the ground.

In just a few million years the two-legged ones
learned to make camps in safe, sheltered places.
They began gathering food in bags made from bark
and hunting animals with spears and sharp stones.
Then they brought back their food to cook on the fire
and to share around with each other.
At night in the firelight they talked, danced and sang,
and told stories of long ago rather like this one.
Whispering, wondering and laughing together
they were the very first people.
And we are their great, great, great, great . . .
                                    . . . great grandchildren.

First edition for the United States the Philippines, and
Canada published 1988 by Barron's Educational Series, Inc.

Life Story was conceived, edited and designed by
Frances Lincoln Limited, Apollo Works,
5 Charlton Kings Road, London, England.

All inquiries should be addressed to:
Barron's Educational Series, Inc.
250 Wireless Boulevard
Hauppauge, New York 11788

International Standard Book No. 0-8120-5941-7

Library of Congress Catalog Card No. 87-73253

Printed and bound in Hong Kong

Design and art direction Debbie MacKinnon
Editor Kathy Henderson